JUL 1 5 2008
J 577.16 SCA

WITHDRAWN

20 Questions: Science

What Do You Know About Food Chains and Food Webs?

PowerKiDS press
New York

Suzanne Slade

To Ms. Janet Wieand, who never runs out of energy and is one of the most enthusiastic book lovers I know

Published in 2008 by The Rosen Publishing Group, Inc.
29 East 21st Street, New York, NY 10010

Copyright © 2008 by The Rosen Publishing Group, Inc.

All rights reserved. No part of this book may be reproduced in any form without permission in writing from the publisher, except by a reviewer.

First Edition

Editor: Amelie von Zumbusch
Book Design: Kate Laczynski
Photo Researcher: Jessica Gerweck

Photo Credits: Cover, p. 1 © www.istockphoto.com/Sandra vom Stein; p. 5, 8–10, 13, 21 Shutterstock.com; p. 6 Clip Art; p. 7, 11(bottom), 22 © www.istockphoto.com; p. 11(top) © www.istockphoto.com/Nico Smit; p.12 © www.istockphoto.com/Michael Price; p. 14 © www.istockphoto.com/Georgia Davey; p. 15 © AnimalsAnimals; p. 16 © Mike Parry/Getty Images, Inc.; p. 17 © www.istockphoto.com/David T. Gomez; p. 17 Clip Art; p. 20 © www.istockphoto.com/Tracy Tucker.

Library of Congress Cataloging-in-Publication Data

Slade, Suzanne.
  What do you know about food chains and food webs? / Suzanne Slade. — 1st ed.
      p. cm. — (20 questions : Science)
  Includes index.
  ISBN 978-1-4042-4202-9 (library binding)
  1. Food chains (Ecology)—Miscellanea—Juvenile literature. I. Title.
  QH541.14.S56 2008
  577'.16—dc22

2007035063

Manufactured in the United States of America

Contents

Food Chains and Food Webs 4
1. What might a food chain look like? 6
2. How does a food chain start? 7
3. What makes up a food chain? 7
4. What kinds of animals eat only plants? 8
5. How do secondary consumers catch and eat other animals? 9
6. Can any animals fight off predators? 10
7. How else do animals stay safe from predators? 10
8. Are some predators simply faster than their prey? 11
9. What predators work together to hunt food? 12
10. Do any predators set traps for their prey? 12
11. Do animals ever just steal food from other animals? 13
12. What happens to dead plants and animals that are not eaten
    by scavengers? 14
13. Are there food chains in oceans and lakes, too? 15
14. What is an example of a food chain in the ocean? 16
15. How long is a food chain? 17
16. Do living things belong to more than one food chain? 17
17. How do food chains make a food web? 18
18. Are people part of food webs and food chains? 20
19. Does a food chain always stay the same? 21
20. Do people change food chains? 22
Glossary 23
Index and Web Sites 24

Food Chains and Food Webs

Every living thing needs **energy** to live and grow. Animals and plants get their energy from food. Most plants use sunlight, water, and air to make their own food. Animals cannot make food. Many animals eat plants and get energy from the food the plants make. Other animals feed on plant-eating animals and get energy from plants in this way.

Plants and animals are **connected** to each other by this movement of food energy. This is called a food chain. In any one **habitat**, there might be several food chains. When these food chains connect together, they form a food web.

This little chipmunk is eating a berry. Chipmunks are small, furry animals that live in North America. They eat many foods, such as seeds, berries, and worms.

1. What might a food chain look like?

One example of a food chain starts with a sunflower plant that makes tasty sunflower seeds. A bird eats these seeds. Later, a hungry fox eats the bird. The sunflower, bird, and fox make up this simple food chain.

The arrows in the food chain above show how energy moves through the chain. A cardinal gets energy when it eats sunflower seeds. When a fox eats a cardinal, it gets energy from that bird.

2. How does a food chain start?

Plants, such as these corn plants, depend on their green leaves to make energy for them.

Food chains begin with plants because plants make their own food. Plants do this using a method called **photosynthesis**. Photosynthesis requires sunlight, water, and gas from the air. Since they can produce food from these things, plants are called producers.

3. What makes up a food chain?

A food chain has both producers and **consumers**. Primary consumers are animals that eat producers. Animals that eat plant-eating animals are called secondary consumers.

4. What kinds of animals eat only plants?

Animals that eat plants are **herbivores**. Many different kinds of animals are herbivores. For example, butterflies drink nectar, or sweet juice inside flowers. Birds often eat seeds, nuts, and fruit. Larger animals can also be plant eaters. Elephants eat leaves, roots, grasses, and fruits, while zebras eat mostly grass.

Hummingbirds get most of their food from flowers. These small birds drink nectar from flowers.

5. How do secondary consumers catch and eat other animals?

Animals that hunt and kill other animals are called predators. The animals predators eat are called prey. Predators have special ways of catching their prey. For example, owls have sharp eyesight for finding small animals in the dark. A coyote's excellent sense of smell helps it notice mice in the grass. Lions and wolves have sharp teeth that can tear up meat.

Lions eat mostly large prey, such as zebras and wildebeests. If these large animals are hard to find, lions will also hunt smaller animals.

6. Can any animals fight off predators?

The barbs on this blue-spotted stingray's tail hold venom, or poison.

Many animals have built-in **weapons**, such as claws and teeth. A porcupine has about 30,000 pointed quills to fight off enemies. The stingray has sharp barbs in its tail.

7. How else do animals stay safe from predators?

Some animals hide from predators by looking like their surroundings. This is called **camouflage**. A white-tailed deer's brown and gray fur helps the deer hide while standing among trees or lying on the forest floor.

Hyenas also catch prey by running fast. A hyena can run faster than 35 miles per hour (56 km/h).

8. Are some predators simply faster than their prey?

Yes, several predators use speed to catch prey. For example, cheetahs and lions can run faster than most of their prey.

When an enemy appears, small animals, such as mice, chipmunks, and groundhogs, often hide in their underground homes. Larger animals may stay safe by running away.

This hare's brown coat camouflages it. It is hard to see the hare against the grasses and dirt behind it.

9. What predators work together to hunt food?

Gray wolves often hunt in groups called packs. The wolves **howl** to tell each other where to gather for a hunt. Wolves look for large prey such as elk, moose, and deer.

Lions also hunt in groups. One group member usually kills the prey, but many lions enjoy the freshly caught feast together.

10. Do any predators set traps for their prey?

Spiders spin sticky webs that trap many kinds of bugs.

African hunting dogs hunt in packs, just as wolves do. Most African hunting dog packs have between 6 and 20 members, though some have as many as 40 members.

11. Do animals ever just steal food from other animals?

Though the silk spiders spin their webs out of is very thin, it is also very strong. This keeps the spider's prey from breaking out of the web.

Scavengers are animals that eat prey another predator has killed. These meat stealers also eat animals that have died from natural causes. One scavenger is a large bird called a vulture. It eats animal remains predators have left behind and animals that have been hit by cars.

Vultures have bare heads. This helps them stay clean when they are eating the bodies of dead animals.

12. What happens to dead plants and animals that are not eaten by scavengers?

The remains of many dead plants and animals are eaten by bugs, **bacteria**, or **fungi**. These feeders are called decomposers. They help decompose, or break down, dead plants and animals and turn them into new, rich soil.

Mushrooms, like these ones growing on a dead tree, are one kind of fungi. Like most fungi, these mushrooms are decomposers. Some mushrooms are tasty, while others are poisonous.

13. Are there food chains in oceans and lakes, too?

Food chains are found in every habitat. A habitat is a place that supplies food, water, and homes for the animals that live there. Oceans, lakes, forests, mountains, and deserts are examples of habitats. The food chains in oceans and lakes begin with producers, such as plants and **algae**, that grow in water.

Northern pike live in the lakes and rivers of North America, Asia, and Europe. Pike are predators at the top of the food chain. These fish generally hunt by hiding behind water plants and springing out when prey comes along.

14. What is an example of a food chain in the ocean?

Tiny living things called **phytoplankton** are the main producers in the ocean. Small **zooplankton** feed on phytoplankton. Fish, such as herring, eat zooplankton. These fish make a great meal for tuna and other larger consumers. Fishermen often catch tuna. Some tuna is canned to sell in stores. If you eat a tuna sandwich, you are another consumer in this food chain that began in the ocean.

Great white sharks, like this one, eat many kinds of smaller fish. This big shark has no natural predators. This makes it the apex, or top, predator in the food chain.

The food chain that this elephant is part of is a very short one. It has just a producer and a single consumer. Elephants eat plants, but no animals eat adult elephants.

15. How long is a food chain?

Some food chains are short. They start with a plant and have two or three consumers. Other, longer chains have many consumers.

16. Do living things belong to more than one food chain?

Yes, most plants and animals are part of several food chains.

17. How do food chains make a food web?

One animal can be part of many food chains because it eats several different kinds of food. Different animals may hunt the same one animal, too, making that animal part of several other food chains. Therefore, one animal connects many food chains in a habitat into one food web.

For example, a mouse eats many different plants. Several secondary consumers, such as owls, snakes, and raccoons, may eat this mouse. The plants this mouse eats and the predators that eat the mouse are all part of different food chains. All these food chains are connected into one large food web by the mouse.

This food web shows the way several living things in one habitat are tied together. The arrows leading away from a plant or animal point to the consumers that eat it.

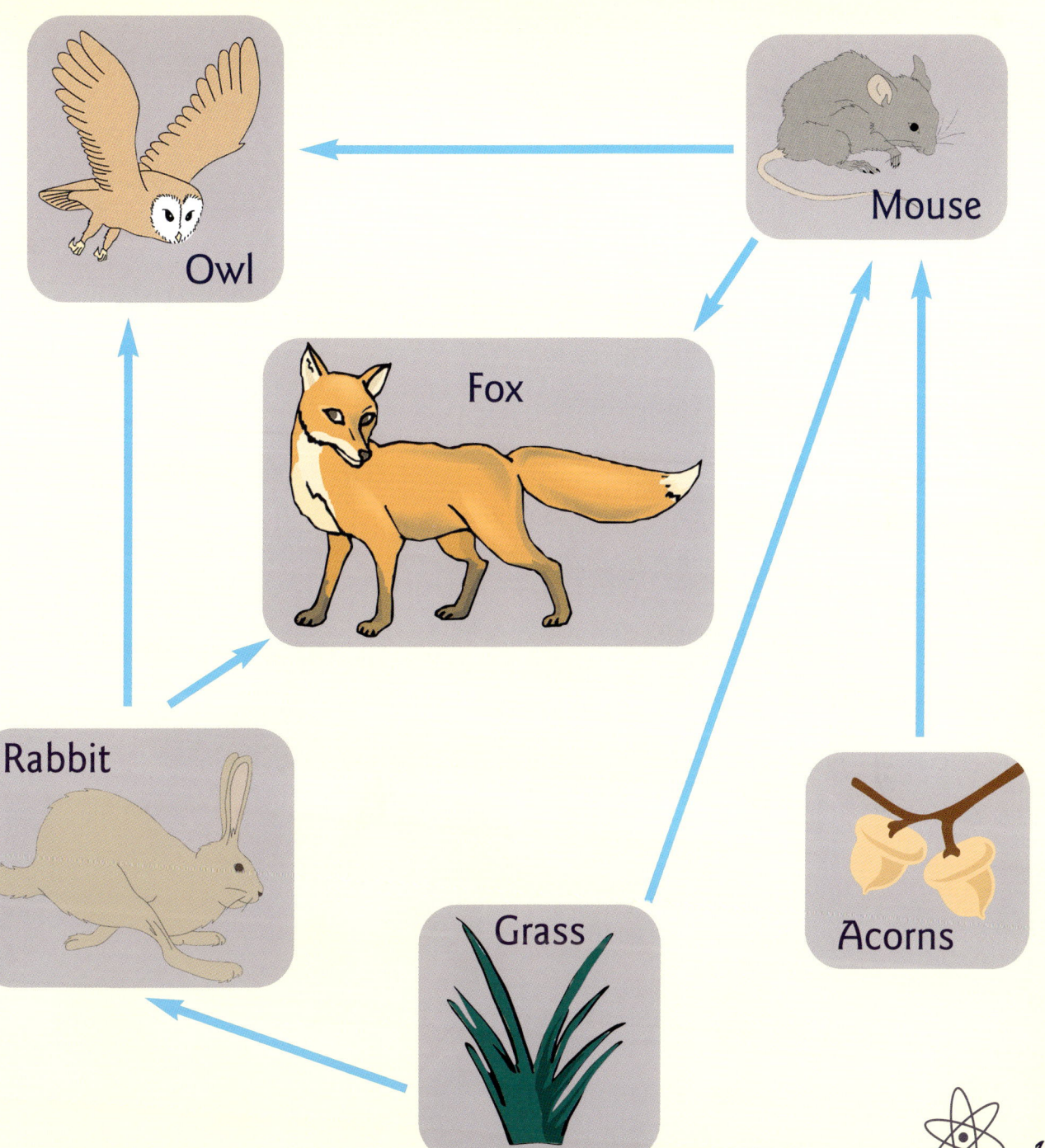

18. Are people part of food webs and food chains?

People are consumers of many kinds of plants and animals. Fields all over the world grow different crops to supply the plants people eat. Apples, oranges, carrots, corn, beans, and the flour people use to make bread all come from plants. In some places, people hunt for meat, such as deer and turkey. In other places, farmers raise cows, pigs, chickens, and other animals for people to eat.

The many kinds of fruit people eat, such as watermelon, all come from plants. Fruits are good for you. People should eat several servings of fruit each day.

19. Does a food chain always stay the same?

Food chains change because the plants and animals that live in a habitat change. Some plants or animals may die out, or become extinct. When a living thing becomes extinct, all the food chains it belongs to are changed.

In 2004, the waters in the North Sea became warmer. This caused the number of zooplankton to go down, which led to a smaller number of the fish, such as sandeels, that eat zooplankton. The birds, such as guillemots, that feed on these fish were hurt, too. Most years, thousands of guillemots are born on Scotland's Fair Isle. In 2004, no guillemots lived to be adults there.

20. Do people change food chains?

Over six and a half **billion** people live on Earth. Our growing population needs lots of land for homes, businesses, and farms. People sometimes cut down forests or cover wet land with dirt to make new places for buildings and farms. Cars and factories also **pollute** the land, air, and water. Without meaning to, people often change food chains when they destroy habitats or hurt producers and consumers with pollution. Today, people are working to find new ways to meet our growing needs while keeping the plants and animals in our food chains safe.

Each year, people burn hundreds of square miles (sq km) of South America's Amazon rain forest to make more room for farms. This destroys the rain forest's producers and hurts its natural food webs.

Glossary

algae (AL-jee) Plantlike living things without stems.

bacteria (bak-TIR-ee-uh) Tiny living things that cannot be seen with the eye alone.

billion (BIL-yun) One thousand millions.

camouflage (KA-muh-flahj) Hiding by looking like the things around one.

connected (kuh-NEKT-ed) Tied to or having to do with.

consumers (kun-SOO-merz) Members of the food chain that eat other living things.

energy (EH-nur-jee) The power to work or to act.

fungi (FUN-jy) Plantlike living things that do not have leaves, flowers, or green color, and that do not make their own food.

habitat (HA-beh-tat) The kind of land where an animal or a plant naturally lives.

herbivores (ER-buh-vorz) Animals that eat plants.

howl (HOWL) To make a loud, long cry.

photosynthesis (foh-toh-SIN-thuh-sus) The way in which green plants make their own food from sunlight, water, and a gas called carbon dioxide.

phytoplankton (fy-toh-PLANK-tun) Ocean producers made up of one cell.

pollute (puh-LOOT) To hurt with harmful matter.

weapons (WEH-punz) Objects used to hurt, kill, or scare away.

zooplankton (zoh-uh-PLANK-tun) Tiny consumers or decomposers that float freely in water.

Index

A
algae, 15

B
bacteria, 14

C
camouflage, 10
cars, 13, 22
consumers, 7,
    16–18, 20, 22

E
Earth, 22
energy, 4

F
fungi, 14

H
habitat(s), 4, 15,
    21–22
herbivores, 8

P
photosynthesis, 7
phytoplankton, 16

W
weapons, 10
wolves, 9, 12

Z
zooplankton, 16

Web Sites

Due to the changing nature of Internet links, PowerKids Press has developed an online list of Web sites related to the subject of this book. This site is updated regularly. Please use this link to access the list:
www.powerkidslinks.com/20sci/chain/